W9-BEB-541

Eco-Camping

Published by Abdo & Daughters, 4940 Viking Drive, Suite 622, Edina, Minnesota 55435.

Library bound edition distributed by Rockbottom Books, Pentagon Tower, P.O. Box 36036, Minneapolis, Minnesota 55435.

Copyright © 1993 by Abdo Consulting Group, Inc., Pentagon Tower, P.O. Box 36036, Minneapolis, Minnesota 55435. International copyrights reserved in all countries. No part of this book may be reproduced in any form without written permission from the publisher.

Printed in the United States.

Compiled by Bob Italia
Illustration (p. 7) courtesy of David J. Taft, National Park Service Ranger
Illustration (p. 11) by Craig Parent

Cover Photo: Peter Arnold, Inc.
Inside photos courtesy of the National Park Service.

Library of Congress Cataloging-in-Publication Data

Italia, Robert, 1955-
    Eco-camping/Bob Italia
       p. cm. -- (Target earth)
    Includes index.
    ISBN: 1-56239-206-9
    1. School camps--United States. 2. Camping--Environmental aspects--United States.
I. Title. II. Series.
GV 197.S3I83 1993
371.3'8--dc20   93-10799
                                                      CIP
                                                      AC

 Thanks to the trees from which this recycled paper was first made.

Table of Contents

Editor's Note

Gateway National Recreation Area was one of the first two National Recreation areas established in an urban metropolis. The park's 26,000 acres in New York and New Jersey border New York Harbor, forming a gateway to the city. With more than 6 million visitors each year, it is the fifth most visited National Park in America.

The overnight tent camping program at Gateway National Recreation Area's Jamaica Bay unit in Brooklyn, New York, began in 1977. The program was inspired by the creative leadership of Samuel Holmes, former Chief of Interpretation and Recreation at Gateway, and Dr. Eugene Ezersky, former Assistant Director of Outdoor Education and Camping for the New York City Board of Education. These men shared the belief that urban children benefit from an experience that combines lessons in ecology with cooperative living and problem-solving.

Much of the "how-to" material in this book is the work of an exemplary teacher and camper, David L. Cohen, President of Educators For Gateway. This material grew out of nearly 15 years of the staff development course offered at Gateway Environmental Study Center. Active participation by interested teacher/campers has been and always will be essential to the success of camping at Gateway.

Special thanks to Maggie Zadorozny, Environmental Education Specialist at Gateway, for her invaluable help and cooperation.

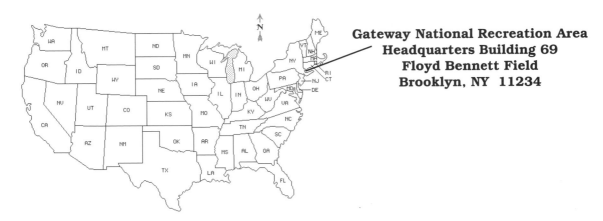

Gateway National Recreation Area
Headquarters Building 69
Floyd Bennett Field
Brooklyn, NY 11234

A Word About
the National Park Service

The National Park Service is a bureau of the U.S. Department of the Interior. The "Organic Act" of 1916 established the National Park Service ". . . to conserve the scenery and the natural objects and the wildlife therein and to provide for the enjoyment of the same in such manner and by means as will leave them unimpaired for the enjoyment of future generations. . ."

It is the mission of the National Park Service to protect and maintain these cultural and natural resources for the use and enjoyment of people today and for generations to come.

There are 367 sites in the system nationwide. These sites include national parks, historic sites, recreation areas, preserves, seashores, scenic rivers, lakeshores, and scenic trails. The Park System is divided into 10 regions. Gateway National Recreation Area is part of the North Atlantic Region which consists of 40 sites. For further information about National Parks in your region, write to:

U.S. Department of the Interior
P.O. Box 37127
Washington, D.C. 20013-7127.

What's Eco-Camping?

There is nothing quite like going to sleep surrounded by wild dunes, pine forests, uplands, coasts, hills, or mountains and waking up to the sound of birds calling. Camping is great fun. You can get close to nature as you learn about yourself and the environment. But with more and more people camping these days, how can you camp without disturbing the environment? That's what this book is all about.

Ecology describes the way plants and animals interact with their environment. Eco-camping helps to show these relationships in a number of ways. You can see for yourself how human activities affect the natural environment when you observe other camp sites and discuss the impact on life there. You can learn how plants and animals survive in the area. You can discover how you affect the environment and how you interact with others by spending the night in a situation that requires everyone to cooperate.

A teacher or adult supervisor is an important part of this experience. Even the toughest "street kid" will feel apprehensive in the open spaces, the dark, the unfamiliarity of tents and outdoor living. An adult can give the support and confidence young people need to enjoy and benefit from the camping trip.

Planning Your Eco-Camping Trip

Be sure to carefully prepare for your eco-camping trip. All the things you can do beforehand or in the classroom will add to the anticipation of the trip. The ideas included here are some ways you might combine camping with learning.

Pre-Trip Activities

1. Write a letter to parents describing the trip.

2. Write the letter requesting parent volunteers and describing the job.

3. Use the library or consult a school dietician for advice on planning nutritionally-balanced meals for camping. Take note of recommended sizes of servings per person and suggested preparation techniques. (For more information, see Chapter 5—Eco-Menu.)

4. Assemble tentative menus and lists of nonfood, condiments, and other supplies, then comparison shop. You should look for and record price differences between name brands, house brands, and generic items.

Responsible eco-camping starts in the classroom.

5. Use the library or consult managers of produce and meat departments in a supermarket to find out what foods are in season or may be on sale during the time of the camping trip.

6. Create a class store and do a mock shopping to emphasize the need for making environmentally sound purchases.

To set up the store, collect empty food and paper good containers of a wide variety:

- recyclable and non-recyclable plastics
- plastic fruits and vegetables, some wrapped in a styrofoam tray covered in plastic wrap and others loose to represent how real produce is marketed
- individual juice containers and juice in larger, recyclable containers

- paper plates and cups and reusable plastic plates and cups
- paper towels, sponges, and cloth towels
- large economy-size cereal boxes and single-serving boxes
- any other likely food or nonfood items you might find in a store.

Shop in teams for the camping trip. Discuss each team's selections in terms of its impact on the environment. Talk about trash disposal, landfill management and using natural resources wisely. (Styrofoam is a petroleum product that is not easily recycled, cannot be washed and reused, and does not break down like paper.) Suggest reasons to use cloth towels and sponges rather than paper towels. (A juice pack is made of as many as six layers of paper and plastic and cannot be recycled like simpler paper products.) Have the teams determine ways they can make environmentally sound purchases for a camping trip.

7. Learn from the other Target Earth™ Earthmobile books about the food chain, web of life, cycles of nature (water, air, and mineral), adaptations, pollution, problems created by various kinds of pollution, and the importance of such things like salt marshes. For more information, see *Eco-Solutions: It's in Your Hands, Intro to Your Environment,* and *The Stream Team On Patrol.*

8. Divide your class into groups. Have each group write and prepare a five-minute skit for the campfire based on a story, song, or television program. Each group may also invent or research and teach a new game to the entire group.

9. With your teacher's help, plan to make ongoing, periodic scientific observations during the trip. Prepare charts for recording data and make or secure an appropriate instrument for the observation. These might include:

- weather observations (temperature, wind speed, wind direction, air pressure, cloud formations)

A food chain (short red arrows) is often tied to other food chains to form a complex food web where organisms from different food chains feed off each other.

- ♦ astronomical observations (recording the number of degrees of movement of the sun, moon, or a star in a constellation at period intervals)

- ♦ using star charts to develop a list of constellations to look for on the horizon at any time on the evening of the camping trip)

- ♦ look for and draw or photograph plants and animals in the campsite area

10. Plan fund-raising activities to pay for the trip. For more information, see the Target Earth Earthmobile book *Eco-Fairs and Carnivals.*

11. Read about gardening.

12. Select and prepare songs for the campfire. They can be taught in class.

13. Develop safety rules. Prepare a safety contract. Create camping safety brochures and posters and safety skits to be performed in school or at the campfire.

14. Prepare a detailed schedule for the trip with times and activities listed. Include a specific list of campfire activities.

You should only buy, bring, and use products and supplies that are environmentally sound. For example, wrap sandwiches in wax paper rather than individual plastic bags. Bring your own reusable plastic cups, dishes and eating utensils or the kits used by Boy Scouts and Girl Scouts so everyone can wash their own. Select large packages of food rather than individual packages. Buy cereals, raisins, and juices in larger units rather than in wasteful single serving package. Use cloth towels rather than rolls of paper for clean up. And pack out as much as you can rather than filling plastic bags with trash to be dumped.

The following camping equipment is recommended:

__basins	__cutting knives
__mixing bowls	__oven mitts
__large pot	__colander
__large pot lids	__cutting boards
__small pot	__propane stove & fuel
__small pot lids	__ground cloth
__frying pans	__ensolite pads
__tongs	__tent brooms
__can openers	__hammers
__peelers	__saws
__cheese graters	__shovels
__spatulas	__coolers
__serving spoons	__water jugs

__backpacks
__two-tined fork
__3-person tents
__ground cloths
__ponchos
__cook set
__first-aid kit
__compasses

__binoculars
__personal gear:
(sleeping bags, clothing, etc.—see Individual Camping Equipment List, page 50).

The physical environment of the campsite you choose can be used to work a profound positive influence on your group. Structure your trip to promote independence and interdependence. Leave civilization, but not civilized behavior, behind. Take pride in real accomplishment. Understand the need for cooperation, and gratitude and respect for the environment.

Camper Orientation

   With your supervisor's help, campers should be assigned a campsite, tents, cookware, utensils, ground pads, coolers, water jugs, etc. Once these items and campers' personal gear are brought to the campsite, the supervisor should hold a camper orientation.

   Part of the orientation, in addition to an explanation of the schedule and rules, is learning cooperation. Every person's cooperation is essential to the success of the camp-out. Without each person's cooperation in certain life sustaining tasks (pitching tents, preparing meals, hauling water), your stay will be less than satisfactory. Do as much for yourself as you are physically, emotionally, and mentally capable of doing.

Chapter 2

Eco-Camping Sites

Most national parks and preserves have campsites. Some, like Gateway, have an established school camping program and provide tents. For the nearest national park camping ground or program near you, contact the U.S. Department of the Interior (see page 6).

Gateway National Recreation Area has an established camping program. Several of these tent-platforms were built by Telephone Pioneers of America volunteers.

Traveling on Trails

Most campers and hikers often use existing trails to reach their campsites. Though traveling on existing trails causes some environmental damage, trails reduce damage to off-trail lands and maintain the solitude that animals need.

When following existing trails, walk in single file and stay on the path. If you travel abreast with a friend, you could break down the trail edge and widen the path. This will promote soil erosion and further damage the environment. You might also trample plants and wild flowers. And you might encourage the formation of new trails which will cut into the off-trail landscape.

Staying on the trail might be difficult if there are muddy stretches. But this is exactly when trails are most vulnerable to damage. When encountering mud, stay on course. If you are wearing comfortable (not heavy or stiff) hiking boots like you should be, the mud will be a slight nuisance.

Another trail problem is encountering approaching hikers and campers. If you do, move off to the side and stop. Walking along the edge will only damage the path. When the other hikers have passed, return to the path and stay on course.

If you need to rest, look for a durable area just off the path. Durable areas include rock outcrops, sandy patches, grassy meadows, or barren patches.

Try to stay on the trail when hiking through the woods.

Sometimes you must travel across open country to reach your campsite. If you do, select a route that avoids fragile areas with vegetation. Most plants will die if stepped upon. And trampling loose soil will cause erosion and prevent vegetation growth. Choose instead a route that crosses rock, gravel, sand, and other durable surfaces. Avoid steep slopes whenever possible.

Hike in groups of four to six people. Spread out instead of following a single path. Don't leave marks or messages in the trees, dirt, or sand. And never hike in open country without the company of an adult. Without exiting trails, it is very easy to get lost in the wilderness.

16

Golden Crow Kinglets.

Respecting Wildlife

When you hike in open country and wooded areas, you enter the home of wild animals like frogs, birds, deer, and bear. Though it is fun to see animals in the wild, people are uninvited guests and can often create unnecessary and harmful disruption of animals' lives. (How would you like it if a group of bears decided to visit you while you were sleeping or playing baseball?)

Don't feed wild animals or leave food lying around. This may make them dependent on human handouts. Don't camp near waterholes where animals come to drink. You may chase them off to a less hospitable habitat. Walking on vegetation disturbs the homes of small animals like insects, mice, birds and squirrels.

Encountering animals on a hike can also create problems. Such encounters often create excitement and alarm in animals.

The white-footed mouse.

Frightened animals may run away and leave their young to die. They may also become confused and drown accidentally in escape attempts.

To reduce your impact on animals in the wild, learn all you can about the ones you might find in the area where you will camp. (If possible, try to avoid bear country altogether. This will lessen your chances for a surprise encounter.) Learn about nesting sites, watering and feeding grounds so you can avoid them. When observing animals, keep your distance and try to stay hidden. And whether you are watching animals or hiking through their homelands, try to keep as quiet as possible.

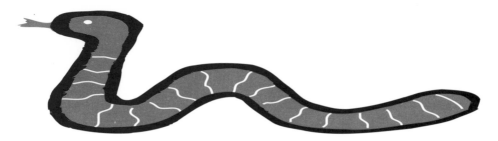

Selecting a Campsite

If you are camping in a national park, most campsites will already be established. Though using the same campsite over and over again creates a lot of damage, overall it spares the rest of the area from harm.

In the backcountry, careful selection is important. You will need level spots for tents and enough room for camp chores and activities, a nearby water supply for washing and cooking, drywood for campfires, and protection from weather.

Dry meadows and open forest clearings with grass make the best campsites. Grass can rebound much quicker than other types of vegetation. A forest or meadow with any other type of ground vegetation are poor choices for campsites. Consider the type of soil and vegetation, and the chance of disturbing wildlife. You'll cause the least damage if you camp on durable sites like rock, dry grass, or existing campsites.

The most harmful damage comes from ground trampling. Trampling soil causes erosion and hardens the remaining soil, making it difficult for young plants to grow. Collecting firewood also creates damage. If you build a campfire, collect drywood. Green, fresh wood does not burn.

An established campsite at Gateway National Recreation Area.

19

Set up tents and kitchen area in places that have already been affected.

Tent Setup

Once you have chosen a camping site, you will need to set up the tents. All tents have different construction and setup procedures. If you follow these general procedures, you will get your camping adventure off to a good start.

1. Place a ground cloth on ground.

2. Unroll the tent on top of the ground cloth.

3. Stake the tent's four corners, making sure the floor is smooth. Work diagonally.

4. Assemble tent poles and place on hardware bag for safekeeping.

5. Insert tent poles.

6. Raise the tent and secure the pole.

Set up tents and kitchen area in places that have already been affected. Paths between these areas should already be established. Watch out for tree seedlings. They are very fragile and not easily replaced.

If you must camp on a new site, spread out the tents and cooking areas. Don't use the same paths between the tents. Move your camp every night if you observe damage to plants or the soil. Wear soft-soled shoes around the campsite. Don't dig a firepit. Use a portable stove (with adult supervision) if possible. If you move rocks for a sleeping area, replace them before leaving.

Fires and Stoves

What would camping be without a campfire? Many people used to think that the campfire was the most important part of camping. But more and more, campers are using portable stoves. Stoves don't harm the environment like campfires do. Carelessly constructed campfires scar the land—and can cause forest fires. They also require a wood supply. That can lead to the destruction of living trees.

Campfires should be built only if a stove is not available. Campfires should be safe, and built in existing campfire sites. These sits should be away from dry grass, trees, branches, and root systems. The best surfaces are bare soil, flat rock, or sparse vegetation.

When building a campfire, scrape or dig a pit several inches deep and make it larger than the area the fire will occupy. Use deadwood for fuel. Collect it from a broad area—don't concentrate your hunt in the immediate campsite. And don't collect more than you need. Don't chop down live trees or break off branches or bark. Build the fire no larger than necessary. Resist the temptation of placing rocks around the campfire. It may look good, but it disturbs the environment. **NEVER LEAVE A CAMPFIRE UNATTENDED.** Try not to build a fire on a windy day or during a drought. Make sure you check fire regulations in your camping area.

Whenever possible, use a stove instead of a campfire.

Campfires should be built only if a stove is not available.
Build your campfire in an existing campfire site.

Chapter 3

Eco-Camping Activities

Now that you're settled in, it's time to have some eco-fun. The following activity involves the food chain and the web of life.

Paste the following little stories on the backs of suitable pictures cut from magazines. (You'll have to do this before you go camping.) Enclose in clear plastic contact paper for permanent use. There is a story for:

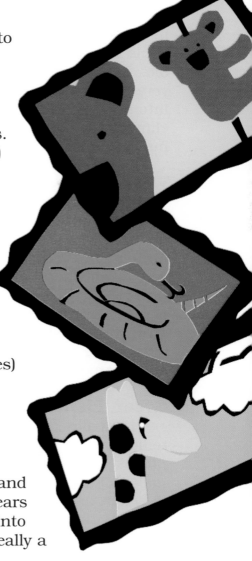

- ◆ the sun (Earth's energy source)
- ◆ the sky (representing air)
- ◆ soil or mountain (representing minerals)
- ◆ the ocean
- ◆ grass, trees and bushes (representing green plants)
- ◆ squirrel or rabbit (representing herbivores)
- ◆ fox or owl (representing carnivores)
- ◆ mushroom (representing fungi which function as decomposers)

Sun Story

I am the source of all energy. I give heat and light to everything. I was formed billions of years ago and will continue sending heat and light into space for many more billions of years. I am really a star.

Air Story

I am made of a mixture of gases. I am made in part of the gas oxygen which all plants and animals need in order to use their food. I am also made of the gas called carbon dioxide, without which green plants could not make food (carry on photosynthesis). I am never used up because, as plants and animals live their lives, they return my parts to the atmosphere.

Mineral Story

I am the different rocks of the Earth. All living things need minerals, either to survive or because parts of their bodies are made of me. For example, without the mineral iron, your blood could not carry oxygen to all parts of your body, and you would die. Without the mineral calcium, your bones and teeth would be very weak. I get into plants because I am dissolved in water. I get into animals when they eat the plants. When plants and animals eliminate waste or die, the decomposers eat the bodies and return me to the soil.

Water Story

All animals and plants are made mostly of water. Take most of the water out of a grape, and you get a raisin! Without me, there is no life. Even though all living things use me, they do not keep me for long, but return me to the environment. Sometimes I am not very clean when I am returned. In the old days, before all these factories and people. I used to be able to clean myself up. Now I just can't work fast enough. I come in three forms: ice, a solid; the form you can slosh or pour, a liquid; and vapor, a gas. I am in the air you are breathing in the form of gas. The form I am in depends on the amount of heat energy I contain.

Green Plant Story

I am alive. So I need enough food, water, and air and the right temperature. I make my own food by doing something called *photosynthesis*. (*Photo* means light, and *synthesis* means to make.) First I take the light of the sun, water, minerals dissolved in water, and carbon dioxide. (Carbon dioxide is a gas given off by all living things as a waste of breathing.) Then in the presence of a green chemical called chlorophyll, I make my own food in my leaves. In making food, I give off a waste gas called oxygen.

24

Squirrel or Rabbit Story

I am alive. So I need enough food, water, and air and the right temperature. I eat green plants. I love tender leaves and nuts. I need to breathe air which contains the gas called oxygen. Without it, the food I eat cannot release its energy. In breathing, I give off a waste gas called carbon dioxide.

Fox or Owl Story

I am alive. So I need enough food, water, air and the right temperature. I eat other animals. My body would not know what to do with plants, even if I tried to eat them. I need to breathe the gas called oxygen. Without it, the food I eat is useless to me. In breathing, I give off a waste gas called carbon dioxide.

Mushroom Story

I am alive. So I need enough food, water, and air and the right temperature. I and my microscopic cousins, the bacteria, are the garbage collectors of the world. We recycle the waste of living things and, after they die, their bodies. What do we do with all the stuff we eat? We put it back so that other living things can use it again. We return minerals to the soil. Together, we fungi and bacteria are called decomposers.

Food Chain Activity

You and your friends (classmates) are going to be creators of a new planet. All of the above eight elements are components for that new planet. You need to add elements so that the planet becomes alive. Which elements should come first?

As you nominate candidates, read the text on the back of each card and decide whether that element can survive. (Hint: the squirrel cannot be the first component because none of its needs can be met).

Web of Life Activity

Place the photographs in a circle on the ground. Which member of the living community can you remove without disturbing the other living members of the newly-created world?

Measure the Size of the Solar System

The Earth is, on the average, 93 million miles (150 million kilometers) from the sun. This distance is called the astronomical unit, or AU. Planets which are closer to the sun are a fraction of one AU. Planets which are further from the sun are more than one AU.

Look at the chart below that shows the average distance of the nine planets from the sun.

Average Distance from the Sun's Radius in Terms of Earth's Distance from the Sun's AU Radius

Mercury	.39
Venus	.75
Earth	1.0
Mars	1.5
Jupiter	5.0
Saturn	9.5
Uranus	9.5
Neptune	30.0
Pluto	39.5

Select one camper to be the pacer. Now find a clearing with a patch of dirt or sand. Draw the sun in the ground with a stick. Position the Earth first by having the pacer take one giant step from the sun. Now position the other planets by having the pacer take the appropriate number of paces from the sun according to the above chart. Draw each planet in the ground with a stick, or have a person stand and represent the planet.

Using the chart and pacing off each planet's distance from the sun, you will get a good idea of the great distances of each planet from the sun. When the activity is finished, erase any lines or marks used to mark a planet.

Chapter 4

Evening Activities

When the adult-supervised activities are over on the first day, it is time to return to camp and begin preparing dinner. Good planning and breaking the group into smaller groups ahead of time will make this run smoothly and be good fun. Everyone can help clean up after the meal, then take a few minutes for themselves to read, talk quietly, write in diaries, or put on a sweater or jacket in anticipation of the evening activities. Then the group reconvenes to listen to the sun set.

Some groups may take a night hike, with no flashlight, to look at the stars and hear the night sounds. Most will start a campfire at sundown. You'll have snacks ready and some activities prepared. As you relax and put aside the work of the day, and night settles on your campsite, you'll notice subtle changes. You will begin to adapt to the unfamiliar physical surroundings and the different social setting. How many have ever been at a campfire before, or sung songs or told stories or roasted marshmallows around it?

You will find yourself becoming closer to your supervisor and your group or classmates. Through a day spent in hands-on activities with the natural environment, you are gaining an appreciation for nature and your relationship to it. Notice how many of you are concerned the next morning about leaving the campsite in good condition!

Because you have had to work together to make the camp run well, you are learning to trust and cooperate with one another on a very basic level. And although some may be apprehensive about the dark, the strangeness and sleeping in tents, you are gaining self-confidence through the experience.

Campfire Activities

If you have a campfire, start the fire shortly before dark. Campers should have prepared an adequate wood pile for the evening. Observe all fire safety precautions.

Activities for Letting Off Steam
(for existing campsites)

1. Scream—Throw a handkerchief in the air. The campers may scream as long as the handkerchief is airborne. Fake throws. Vary the altitude.

2. Hi, My Name Is Bill—This is a spiraling action chant. With every repetition, poor Bill sets another portion of his body in motion to turn yet another machine:

Hi! My name is Bill and I work in a button factory. I have a wife, two kids, and a dog named spot. One day, my boss said, "Bill, turn the buttons with your (right hand, left hand, right/left knee, nose, hips, head. . .)."

3. Honey, I Love You—One student goes over to another, gazes intently and longingly into the other's eyes, and says, with deep emotion, "Honey, I love you."

The other player must respond, without so much as a trace of a smile, "Honey, I love you, but I just can't smile." If this player smiles, laughs, giggles, or shakes, she or he becomes the initiator.

4. Bag of Stuff Story—Set this up in class or during the afternoon before sunset. This assumes the campers have already been divided into groups.

Give each group a similar bag of six of the strangest, most unrelated objects you can assemble. Before the campfire, let each group met in a tent or private place to develop a story which uses all objects. The group members tell or act out the story at the campfire.

5. Sing the Watermelon Song by Bach. Don't forget the slurp.

Oh you can plant a watermelon up above my grave
and let the juice (slurp) seep through.
Oh you can plant a watermelon up above my grave
and let the juice (slurp) seep through.
Oh you can plant a watermelon up above my grave,
that's all I ask of you.
Now a (name a cookie) tastes mighty fine,
but there's nothing quite as good as a watermelon vine.
Oh you can plant a watermelon up above my grave
and let the juice (slurp) seep through.

6. Singing is great if you have time to learn songs in class before the trip. You could also collect songs students already know and prepare a song booklet to bring along.

7. Charades is good, especially for things you saw or did during the trip or people who work in the school.

Away From The Campfire

1. Tangled Web—Students form a circle and reach across to grab a hand. (Even numbered groups of 10 persons or less works best.) Then they reach across and grab the hand of another person. The idea is to untangle the web without anyone letting go.

2. Fox and Goose.—Blindfold two players who stand in the middle of the circle formed by the rest of the group. Fox calls out, "Goose?" and the Goose must answer, "Fox?" Blindfolded Fox tries to locate and tag Goose by sound. Goose keeps moving to evade Fox. Players forming the circle keep these two safe.

Quieter Activities

1. Nature Names—Let students give each other nature names that begin with the same letter as the first letter of the person's real name (e.g., Bobby Berry, Jane Jackrabbit). This can be played in a circle. The teacher can start by naming the camper to his or her right, then that person names the camper to his or her right, and so on around the circle.

2. Memory Mind-Breaker—After everyone has a name, have a contest where a volunteer must remember the nature names of the first five campers, then ten, then all. The prize can be a bag of gorp (see Chapter 5—Eco-Menu) or popcorn.

3. Take a Night Walk. No flashlights.

4. Use star charts to locate constellations and planets.

5. Lie out in a clearing, if it is warm enough. Look at the sky without talking.

6. Scavenger Hunt—This is a good ice-breaker, even for people who think they know each other.

You will have ten minutes to find people who can do the following things. Write the name of the person in the blank. When time is up, please return to your place.

Find someone who:

- Can wiggle his or her ears_____
- Has your middle name_____
- Learned something new today_____
- Has the oldest living grandparent_____
- Was not born in America_____
- Has the strangest pet_____
- Can hold breath longer than you can_____
- Speaks a language other than English_____
  What is that language?_____
- Has one of your hobbies_____
- Has the most unusual allergy_____
- Can imitate the greatest number of animal calls_____
- Has the greatest number of siblings
  (sisters and brothers)_____
- Has two relatives with the same first names
  as two of your relatives_____

31

Chapter 5

Eco-Menu

Good nutrition is a must when camping. As you and your classmates or friends plan the menu for your outing, review the basic food groups and include them in each meal. With the exception of that campfire staple—marshmallows—stay away from junk food. Concentrate instead on tasty alternatives such as fresh or dried fruit, carrot and celery sticks, graham and saltine crackers, bread sticks, peanuts, and popcorn. The same goes for drinks: choose fruit juices (not fruit drinks), milk, or water rather than sodas.

If you are camping for one night, you will have to prepare two lunches, one dinner, and one breakfast. The following recommendations are to help you. Recipes follow. Use your own ideas, too. Try dishes from other cultures, particularly non-meat fare and especially foods prepared at home.

Dinner

One-dish meals are recommended. They are easy, tasty, and nutritious when the right combinations of vegetables, complex carbohydrates, meats, and sauces are combined. Serve carrot sticks, celery, substantial bread and butter, and fresh fruit for a balanced meal.

Serving Sizes

All of the following recipes make 12 large servings. They can be increased or decreased as necessary. Some people may eat more than one serving, especially after a long day outdoors.

Lunches

Try more interesting breads, especially those with whole grain in them. For fillers, use such high-protein foods as peanut butter, cold meats, or tuna salad made just before you eat it. You can make your own peanut butter in a Foley mill or meat grinder. Refrigerate until trip time. Be sure to bring mustard, mayonnaise, ketchup, and other condiments. And don't forget the fresh fruit and raw vegetables.

Breakfast

Instant hot cereals such as oatmeal are a good bet. If you prefer cold cereal, agree with your friends or classmates on one or two types and buy the largest box each. Alternatives to the heavily sugared cereals include, Cheerios, corn flakes, Rice Krispies, and shredded wheat. Have bread, cheese, jelly, and plenty of milk—at least one quart for every 3 or 4 campers. Dry milk powder is good insurance against running out.

Snacks

Stock up on crackers, graham crackers, rice cakes, raisins, fresh fruit, marshmallows, fresh vegetables to eat raw, nuts, popcorn, some cookies, instant hot chocolate, and, for the adults, coffee or tea.

Miscellaneous

You will need salt, pepper, sugar, flour (for thickening gravy), a small amount of oil (for browning or sauteing), butter, ketchup, mustard, mayonnaise, and seasoning for a main dish (chili powder, cumin, parsley, oregano, basil, soy sauce, ginger, tube of tomato paste, etc.).

Use the menu/Shopping List Planner in the Appendix to plan the meals for your trip.

(Sample Recipes)

Chicken and Rice
Preparation time: 1 hour

3 2-3 pound frying chickens, cut up
1/2 c. flour, seasoned with salt and pepper
1/4 c. oil
2 onions, chopped
2 cloves garlic, minced, or 2 tsp. garlic powder
5 tomatoes, chopped or 1 28-oz. can whole tomatoes
1 1-pound package frozen green peas
3 cups uncooked rice
6 cups water

Heat oil in a large pot with a cover. Dredge the chicken in seasoned flour (use a plastic bag and shake). Brown the chicken on all sides, about 15 minutes. Remove from the pan and add garlic powder and onion. Saute 5 minutes. Ad tomatoes, rice and water and arrange chicken on top. Cover the pot and cook 45 minutes or until tender. Add the frozen peas in the last 15 minutes of cooking. Serve with milk and fresh fruit for desert.

Nutrition information: Meat (chicken), vegetables (tomatoes and peas) and grain (rice) are in the main dish. Milk and fruit on the side.

Indian Corn Stew
Preparation time: 1 hour

3 pounds ground beef
2 medium onions, chopped
2 28-oz. cans crushed tomatoes
2 16-oz. cans corn, dry pack
1 large pepper or zucchini, chopped
salt and pepper to taste

In a large pot with cover, brown the ground beef and onion together. Add the tomatoes, corn and zucchini, cover and simmer for 45 minutes. Use salt and pepper if necessary. Serve in bowls with bread and butter on the side, milk and fresh fruit.

Nutrition information: Meat (ground beef), vegetables (tomatoes and zucchini) and grain (corn) are in the main dish. Milk and fruit on the side.

Campfire S'Mores
—24 servings

1 box graham crackers
2 1-pound milk chocolate bars (or 12 7-oz. bars)
1 10-oz. bag large marshmallows
sticks for roasting marshmallows

Give each camper one whole graham cracker, 1 square (or 1/2 bar) of chocolate, and 1 marshmallow. Roast the marshmallows over the fire, then build the s'more as follows:

•Break the graham cracker in two
•Put the square of chocolate on one half
•Put the hot marshmallow on top of the chocolate (carefully)
•Put the other graham cracker on last to make a sandwich

Let it cool a bit.

Gorp
—40 servings

2 pounds raisins
2 pounds unsalted shelled peanuts
1-2 pound plain M&M's or coated carob pieces
Optional: 1 pound shelled sunflower seeds
        1 pound currants

Chapter 6

Breaking Camp

This is the most important task you will complete during your eco-camping trip. Your group should complete all its responsibilities for leaving the campsite as you found it. First, you must take down the tent(s):

1. Put hardware on top of hardware bag to make sure nothing is lost.

2. Sweep out tent.

3. Remove poles and dismantle.

4. Fold the top of the tent to one side.

5. Remove corner stakes on one side.

6. Fold tent in half.

7. Remove corner stakes from other side.

8. Fold tent in fourths.

9. Roll tent towards door.

10. Place poles, stakes, and tents in tent bags.

Now use the following checklist to make sure the campsite is clean:

__campsite totally clean of litter
__stove and campfire dug out and cleaned
__ashes from stove and campfire bagged and deposited in dumpster or trash can (if applicable)
__all garbage bagged and placed in dumpster or trash can (if applicable)
__all trails leading to and from the campsite cleaned of litter

If no trash containers are available, make sure you pack up your trash and carry it out. Don't leave or bury leftover food. This will only encourage animals to forage in camping areas instead of the wilderness.

The Campfire

Blackened campfire sites are ugly scars on the land—and the most obvious sign that campers were around.

To lessen a campfire's impact on the land, avoid a morning fire. If your campfire burned out the night before, the ashes will be cold. You won't need water to douse the ashes and cleanup will be fast and safe.

Crush any charcoal remnants. (If you can see glowing embers, your fire is not out.) Scatter the charcoal remains and any unburned firewood over a large area far from the campsite. Don't dismantle the existing campfire site. Leaving it in tact will encourage others to use it.

Keeping the campsite clean and attractive will make it appealing to other groups. This will encourage them to use your existing campsite instead of creating a new one.

Chapter 7

Post-Trip Activities

You have done something special. You should get the recognition you deserve for it. The Target Earth Earthmobile program deserves recognition, too. Good public relations can ensure the continuation of these programs.

Public Relations

Your public relations audience includes your own school, your parents, your district and your school neighborhood. Student-created materials can effectively demonstrate the integration of language arts and fine arts and can put the "sizzle" into a description of your group's camping experience.

1. Write thank-you letters to any adults who were involved. Describe what you got out of the trip. Don't forget to write to elected officials such as your mayor, members of Congress, and school board members.

Please remember to credit everyone who helped: parent volunteers, colleagues, cafeteria workers, teachers, etc.

2. Set aside a day for teams to report on their scientific investigations.

3. Create drawings, dioramas, models of the campsite, and other memorable elements of the trip for display in school.

4. Write poetry, a short story, or a new camp song about the camping trip and read or perform it for the class or in a school-wide assembly program.

5. Prepare a press release for the school newspaper or newsletter and the local paper. Supply photographs, drawings, names, and other specifics.

6. Follow up on any pre-trip activities such as readings, experiments, observations, and further inquiry.

Appendix—For Adults Only

Pre-Trip Preparation Check List (Sample 2-Day Trip)

1. Describe the trip to the group.

2. Form work groups and select/elect leaders after describing duties of leaders.

3. Provide orientation for leaders on duties including:

♦ safety
♦ morale
♦ cooperation
♦ responsibilities

4. Begin activities (see chapters 3 & 4)

5. Instill respect for rules, hazards, care of equipment, and camper courtesy.

6. Teach songs.

7. Provide time for campers to practice campfire skits/ contributions.

8. Teach new games.

9. Begin to collect materials for a rainy day fall-back (games, crafts activities, and supplies).

10. If this is a school trip, check official health cards for all students.

11. Call parents of all students whose health card indicates a presence of a health problem. Make notes on the health form; include date and parent's instructions.

12. Be certain that someone in a position of authority can be reached by telephone for the entire time of the trip. Stress the importance of having valid telephone numbers.

13. Be certain to have at least one car available at all times.

14. Plan the menu (see Chapter 5)

15. Find out the name and directions to the nearest hospital in the event of an emergency.

Rules for Campers

1. Safety first.

2. Stress a positive attitude. Say that the phrase for the trip is "Yes, I will" or "Yes, we can do it."

3. No private food or candy after lunch is over.

4. No radios or tape players.

5. Necessities before luxuries: work before play.

6. Pay attention when directions are given.

7. Follow directions exactly. Ask questions if you're not sure.

8. No one moves until the entire group is ready.

9. Group leader and an adult must know where you are at all times.

10. Go to the bathroom or to get water only in pairs.

11. Visiting in tents is allowed only with permission of all tent mates. Shoes off.

12. No knives or axes.

13. If you break something, you are responsible for replacing it.

14. Out of bed penalty: sleep in adult tent.

15. Notify adult if going to the bathroom after lights out.

Special Camping Hazards

1. Avoid ticks; stay in the middle of trails; cover up; wear light-colored clothing; have frequent tick checks; and check bedding before lights out.

2. Go to the waterfront/lakeshore only with adults.

3. Stoves and lanterns get hot and will cause burns.

4. You can fall or stumble into a fire or hot stove or other hazard. Do not run, wrestle, or push in the campsite.

5. Use cooking knives carefully when preparing food, and use mitts to handle hot pots.

6. Rinse cooking equipment well.

7. Use sun screen, wear hats, dress in layers, drink water, bring rain gear and stay dry.

8. Adults douse the campfire and use a shovel to turn and inspect ashes. Keep water handy.

9. Wear well-fitting shoes. Pull socks tight.

Camping Courtesy

1. Pick it up if nature did not put it there, even if it is not yours.

2. Quiet after lights out.

3. Shovel out the campfire in morning.

4. Police area immediately prior to departure.

5. "Please" and "thank you" are the most important words campers use.

One Week Before the Trip

1. Collect and check all forms.

2. Prepare camper name/task cards.

3. Complete shopping list and begin to shop for nonperishables.

4. Collect a box of kindling.

5. Assemble rainy-day materials.

6. Have campfire activities and songs in place.

7. Have camping activities in place.

8. Confirm transportation.

9. Review hazards, camper courtesy, rules, and personal equipment list.

10. Arrange with school dietician for food donations (if applicable).

11. Draw up and review schedules for sunny and rainy day camping trips. Review with students. Duplicate copies for group leaders, parent volunteers and park staff (if applicable).

12. Confirm parent volunteers (if applicable).

During the Trip

1. Upon arrival at the park or camping ground, find out where the nearest public telephone and first aid station are located.

2. Use the Tent Mate Work Sheet to distribute, check, and collect tenting equipment.

3. Be aware of weather conditions, temperature, and emotional; states of campers (homesickness).

Trip: Day One

1. Take attendance. Make a final check for valid permission and medical forms (if applicable).

2. Leave the trip cards of attending students with school secretary or administer (if applicable).

3. Bring supplies and student packs from classroom to bus assembly area (if applicable).

4. Secure lunches from school dietician (if applicable).

5. Review rules, hazards, and camper courtesy at the campsite. Prepare campers for the next activity on the schedule.

6. Use the Tent Mate Check List to record problems with tents or other equipment, including cooking gear.

7. Adhere to schedule.

8. Remind students that the first morning activity will be packing personal belongings, cleaning out tents, and neatly assembling all supplies and equipment.

Trip: Day Two

1. Use Tent Mate Check List for striking and collecting tents.

2. Do not start a fire on the second (last) day. Plan a cold breakfast, or one which requires hot water if you have a propane stove.

3. Prepare lunch and breakfast simultaneously. Store lunch in personal plastic containers or wax paper. Please use environmentally sound supplies (reusables as much as possible, very little disposable plastic).

4. Leave campsite in the same condition in which you found it; clean the grounds and remove trash.

5. Do not take down tents until their cleanliness and condition have been inspected.

Appendix—Forms

The following sample forms should be used if the overnight camping trip involves a school or community group.

Overnight Camping Trip Participant

Name_____Class_____

Mother_____Father_____

Home Phone_____

Bus. Phone (Mother)_____(Father)_____

Address_____

Camping Trip Dates_____

Parent Information Sheet

Dear parents or guardian:

We are planning an overnight camping trip to_____.
Our teacher/leader_____has prepared
a camping program based on good camping practices. In addition,
we will have other adults on the trip to help supervise. I hope that
you will allow your daughter/son to participate.

You will find additional information below.

Very truly yours,

_____, Principal

Dates and Times: Campers report to the_____
by_____ A.M./P.M. and depart by____A.M./P.M. the following day.

Packs: Campers must have their packs in school by_____.
The packs will be locked in a secure closet overnight.

Dismissal: We will return between_____and_____on_____.
Students will be dismissed at the normal time unless a parent, or
adult with written permission, is there to pick up the student.

Location:_____

Cost (if applicable): $_____will cover the cost of dinner,
breakfast, lunch, campfire treats and incidental group expenses.
Please do not allow the student to bring candy or extra food on the
trip.

Activities: All activities are conducted by teachers/adult
supervisors. They include nature studies, ecology, hiking,
gardening, and exploration. Students will learn such camping
skills as pitching tents, preparing food outdoors, and working in
groups to get jobs done. We will be outdoors for the majority of this
trip, provided the weather is favorable.

Sleeping Arrangements: Students will sleep in unheated, open-
front tents.

Student Behavior: It is essential that all students behave in a
responsible manner on this trip. In giving permission, you are also
agreeing to be available to pick up your student at any time should
she or he behave in a way which endangers the safety or well-being
of anyone.

Supervision: The teacher/leader and parent volunteers supervise
the group.

Emergencies: In the unlikely event of an emergency, we will contact you at the telephone number you have provided. Please be certain that we can reach you by phone the entire time, as hospitals will not generally treat other than extreme emergencies without a parent or legal guardian present.

Equipment: A personal equipment and clothing list is attached. Please adhere to this list, especially concerning layers of clothing, footwear, and rain gear. If you do not have a warm sleeping bag, please supply three warm blankets.

Please: No radios, tape players, electronic games, knives, or axes.

I can help by (check one):

__Serving as a parent volunteer on the trip.
__Helping with the shopping.
__Driving supplies to the campsite.
__Lend camping equipment (please specify)_____

Name_____Phone_____

Medical Information and Permission Form

I give permission for (student)_____
(address)_____
(phone)_____to attend an overnight camping
experience at_____
on (Day 1)_____ and (Day 2)_____.

Signed:_____ Date:_____
             (Parent or Guardian)

Please complete the following:

Student's name_____
Parent/Guardian's name_____
Home Address_____
Home Phone_____Business phone_____
Family doctor_____Phone_____
Any medication currently taking_____
Last tetanus shot_____
Any health factors which make it advisable to limit physical
activities_____
Any other current medical condition_____

In case of emergency, and you are not home, whom should we
contact?

Name_____Phone_____
Relationship to student_____
Comments:

Individual Camping Equipment List

__Backpack or soft luggage (you may also securely wrap and tie personal belongings in a blanket)
__Sheet (optional)
__Three-season sleeping bag OR three warm blankets
__Towel and wash cloth
__Washable plastic cup
__Washable plastic plate (no styrofoam or plastic-coated paper)
__Washable plastic bowl or canister for cereal
__Fork, spoon, and a dull knife for meals
__Toothbrush and paste
__Soap
__Comb and/or brush
__Two sweaters and a sweatshirt
__One long-sleeve flannel shirt
__Good walking shoes (no open shoes or sandals)
__Two pair thick socks
__One full change of clothing
__Raincoat or poncho, hat, and waterproof boots
__Laundry bag
__Sun screen and sun hat
__Pad and pen
__Ground pad

For cool weather:
__Warm jacket
__Gloves
__Warm hat and scarf

Optional:
__Flashlight
__Board game
__Book
__Camera
__Ball
__Binoculars
__Insect repellant
__Canteen

Teacher/Leader:
__Class list
__Medical form for each student
__Complete Work Assignment Sheet
__Matches
__Camp stove and fuel
__Toilet paper
__Knife
__Can opener
__Saw or axes
__Lantern (optional)
__All food
__Ice and coolers
__Tents and ground cloths
__Pot Holders
__Pot scrubbers and dish soap
__Plates, cups, eating utensils
__Sun screen
__Group First Aid kit
__Misc. (see page 33)
__Pots and pans
__Cooking utensils and cutting board
__Water jugs
__Hammer (optional)

No electronic games, radios, or tape players

Please detach the following and return with signed medical information and permission form.

I have read and understand the information about equipment and the ban on electronic devices.

_____   _____

Parent/Guardian's name   Date

Forms Check List

You will need to complete several forms for the trip. You will also need to check and/or collect several forms or items from your students. Duplicate this check list to help you keep track of this paperwork.

General forms:

1. Class trip permission
2. Transportation request
3. Transportation confirmation
4. Individual forms: Parent Info and Medical Form; 3 x 5 trip card; permission slip.

Name	Parent Info.	Medical Form	Trip Card	Permission Slip

Tent Mate Check List

Fill out this checklist with the names of tent mates and bring it with you.

Tent mates	tent #	problems	clean	tally

After the supervisor has inspected each tent and noted any problems beside those noted at the setup, campers are to strike and fold tents as demonstrated. Students should lay out stakes, poles, tent, stake, and tent bags and ground cloths. After you tally all equipment, campers will pack it up.

Menu/Shopping List Planner

See Chapter 5 for suggested recipes or use your own. Then use this form to plan your menu and shopping list. Take into account ease of preparation, storage, and spoilage; ease of cleanup; cost (look for seasonal bargains); nutritional value and balance; amount of preparation and cooking time; and time of sunset (ideally, you want to be completing cleanup just as the sun is setting).

You will plan for three meals and a campfire snack.

Meal	Food	Condiments
Dinner		
Campfire		
Breakfast: Day 2		
Lunch: Day 2		
Nonfood items:		

Work Assignments

Before you arrive, make group assignments using the class list below. Then assign work groups on the camping schedules that follow. Please make photocopies of this schedule for all adult supervisors.

Class List/Group Assignments

Divide your class into 5 groups of 6 students each (ideally, 3 girls and 3 boys in each group). Write the name of each student in the appropriate groups. Use appropriate names for each group. For example:

Raccoons

1.
2.
3.
4.
5.
6.*

*student leader

Camping Schedule
(sample)

Day 1

10:00 A.M.	Check in at campsite. Deposit gear.
10:30	Orientation
11:00	Camp Set Up
	_____Group—Set up kitchen
	_____Group—Fill water jugs
12:00 P.M.	_____Group—Lunch preparation
	Camp lunch—all groups
	_____Group—Lunch cleanup
1:00	Nature hike
4:00	_____Group—Dinner preparation
	Camp dinner—all groups
	_____Group—Dinner cleanup
7:00	Campfire
10:00	Lights Out! Extinguish campfire

Day 2

7:00 A.M.	Wake up. Cold breakfast
	_____Group—Breakfast preparation
	_____Group—Breakfast cleanup
8:00	Break camp. Tent inspection
	_____Group—Load equipment
	_____Group—Inspect campsite for articles/litter
9:00	Group activity
11:00	_____Group—Lunch preparation
	_____Group—Lunch cleanup
12:00 P.M.	Depart for school

Emergency Procedures

In case of an emergency, remain with your group and send another adult for help. If you are not in a park, take the injured camper to the nearest emergency room. Make sure to bring along the injured camper's medical form. Once at the hospital, telephone the parent or guardian.

WALKERTON ELEMENTARY
SCHOOL LIBRARY

Glossary

Acid rain—when harmful gases from cars and factories are released into the air and fall back to the Earth with rain or snow.

Algae—tiny plants that lack stems, roots, and leaves. Usually found in water.

Atmosphere—the layer of gases surrounding the Earth; the air.

Carbon dioxide—a gas produced when animals breathe out or any material containing carbon is burned.

Carbon monoxide—a poisonous gas released when wood, coal, oil, gasoline, natural gas and other fuels are burned.

Carnivore—an animal that eats the flesh of another animal.

Decomposers—bacteria, molds, and fungi that devour plant matter.

Drought—a long, dry period of weather, with little or no rain.

Ecology—the study of the interrelationships between organisms and their environment

Ecosystem—the interaction of plants, animals, and other natural elements in an interrelated system.

Environment—all the living and nonliving surroundings of an organism.

Environmental science—the study of interrelationships between organisms and their environment, with a major focus on humans.

Exploitation—the overuse, misuse, waste and destruction of natural resources.

Extinction—the end of an organism's existence on Earth.

Exotic species—plants or animals that are introduced to an ecosystem from distant places and create a new set of problems.

Food chain—the flow of nutrients and energy among a series of organisms that feed on each other.

Food web—a connecting series of food chains.

Fossil fuels—fuels like coal, oil and natural gas that were produced millions of year ago from dead plants and animals.

Greenhouse effect—when gases from factories and cars trap the sun's heat and warm up the Earth.

Groundwater—water that has seeped into the soil and collected in underground spaces.

Habitat—an area that provides an animal or plant with food, water, shelter, and living space.

Herbivore—an animal that feeds on vegetation only.

Landfill—a system of rubbish disposal in which the solid waste is covered with soil.

Natural resources—any part of the world of nature that has value to humans.

Organism—a plant or animal.

Ozone layer—a layer of gas high in the sky that protects the Earth from harmful ultraviolet (UV) rays of the sun.

Pesticide—a chemical substance used to destroy insects and other pests.

Pollution—the act of dirtying or degrading a natural resource so that it is less valuable to humans.

Preservation—an approach to resource management in which the original character of the resource is maintained.

Primary consumers—animals in a food chain that eat plants. Shredders and herbivores are primary consumers.

Renewable resource—a resource such as soil, water, air, forests, and wildlife, which can be renewed in a somewhat short time.

Science—knowledge based on observations and tested truths.

Secondary consumers—Animals that feed on primary consumers. Small fish and larger insects are secondary consumers.

Shredders—A small group of underwater animals, including snails and insects, that eat decomposers.

Siltation—The filling up of a stream or reservoir with sediment.

Solid waste—rubbish such as cans, bottles, tires, plastics, building materials, and used cloth.

Sulfur dioxide—a colorless gas released when sulfur containing fuels like coal, oil and natural gas are burned.

Third-level consumers—Animals that feed on secondary consumers. Larger fish, small land animals, and humans are third-level consumers.

Watershed—The total area drained by a stream or river.

Index

F

fire—21, 28, 35, 37, 42, 45

firepit—21

firewood—19, 37

first-aid—13

flowers—16

food chain—23, 25

food chain—10

forests—7

forks—13, 50

forms check list—52

fox—25, 30

frogs—17

fruit—32, 33, 34, 35

fund-raising—11

fungi—23, 25

G

gardening—12, 47

gas—24, 25

Gateway National Recreation Area— 5, 6, 15

grass—19, 21

graters—12

ground cloth—12, 13, 20, 51, 53

ground pad—50

H

hammers—12, 51

hikers—15, 16

hills—7

J

jugs—13, 14, 51

juice—9, 10, 29

K

knives—12, 41, 42, 48

L

lanterns—42, 51

lids—12

lunch—32, 33, 41, 44, 45, 47

M

mallets—13

meals—8, 14, 32, 33, 50, 54

medical forms—44

medical information—49, 51

menu—8, 30, 32, 33, 41, 54

menu/shopping list planner—33, 54

mineral—10, 23, 24, 25

mixing bowls—12

moon—11

mountains—7, 23

mushroom—25

N

National Park Service—6

natural resources—6, 10

nature—7, 10, 27, 30, 42, 47

non-recyclable—9

nutrition—8, 32, 34, 35, 54

O

ocean—23

oven mitts—12

owls—23, 25

oxygen—24, 25

P

package—12

pads—12, 14

pans—12, 51

parent information sheet—46

parents—8, 38, 40, 46

paths—15, 16, 21

TARGET EARTH COMMITMENT

At Target, we're committed to the environment. We show this commitment not only through our own internal efforts but also through the programs we sponsor in the communities where we do business.

Our commitment to children and the environment began when we became the Founding International Sponsor for Kids for Saving Earth, a nonprofit environmental organization for kids. We helped launch the program in 1989 and supported its growth to three-quarters of a million club members in just three years.

Our commitment to children's environmental education led to the development of an environmental curriculum called Target Earth™, aimed at getting kids involved in their education and in their world.

In addition, we worked with Abdo & Daughters Publishing to develop the Target Earth Earthmobile, an environmental science library on wheels that can be used in libraries, or rolled from classroom to classroom.

Target believes that the children are our future and the future of our planet. Through education, they will save the world!

Minneapolis-based Target Stores is an upscale discount department store chain of 517 stores in 33 states coast-to-coast, and is the largest division of Dayton Hudson Corporation, one of the nation's leading retailers.